I0485287

# COMPETITIVE BIOLOGY 2

## INTRODUCTION

This objective biology series provides a basic and challenging problem of biology from particular topics. It can be used to brush up ones basics and checking up the preparation level of particular topic. It is equally helpful to the traditional classes as well as competitions. It can be also taken as a revision material for any competition which includes the test of basic biology. If you want to grasp the subject before practicing these multiple choice questions, you can go through the website http://www.ncert.nic.in/ncerts/textbook/textbook.htm and down load the free copy of science books and after having command on the topic practice it. For revision purpose, important points are given at the starting of each topic.

If you have any query or suggestion about this series you can send your suggestion at uk2594@gmail.com.

# CONTENTS

# 5. LIFE PROCESS

## SOME IMPORTANT POINTS

➤ All living things perform basic life process like, growth, digestion, excretion, respiration, circulation, etc.

➤ The basic functions perform by living organisms for their survival and body maintenance are called life process.

➤ There are mainly two models nutrition :

1. Autotrophic      2.Heterotrophic

1. Heterotrophic: They synthesis their food by own by the process of photosynthesis.

2. Raw materials required for photosynthesis is $Co_2$ and $H_2O$.

3. Tiny pores present on the surface of the leaves are known as stomata.

4. Stomata performs a special function in the exchange of gases and loses and large amount of water during transpiration.

5. Human alimentary canal originates from the mouth and to the anus.

6. Respiration involves exchange of gases and break down of simple food in order to release energy.

7. Breakdown of glucose by various path ways:

➤ Respiration in plants is simpler than the respiration in animals. Gaseous exchange occurs through.

1. Stomata in leaves

2. Lenticels in stem

3. General surface of the roots

➤ The process of transport food and oxygen, etc. Supply in the multicellular organisms is known as transportation.

➤ Human circulatory system include :

1. Heart

2. Arteries & Veins

3. Blood & Lymph

➤ In human beings blood travels twice through the heart to the body

1. Pulmonary Circulation

2. Systemic Circulation

➤ Blood contains plasma, R.B.C, W.B.C and platelets.

➤ Transportation in plants takes place by xylem and phloem tissue.

➤ Transport of food from leaves to the rest of the body is known as translocation.

➤ The process of removing harmful substances from the body is known as excretion.
➤ Excretory system includes a pair of kidney, a pair of ureter, a urinary bladder and a urethra.
➤ Nephron is the filtration unit of kidney.
➤ The process of purify blood by an artificial kidney is known as Haemodialysis.

## 5. LIFE PROCESS

1.      Name the organic compounds used by plant to make their food?

   (a)Water and minerals       (b)Water and $CO_2$

   (c)Water and Minerals       (d) none of these

2.      Plant converted carbon dioxide and water into in the presence of

   Sunlight & chlorophyll?

   (a)Proteins              (b) Carbon monoxide

   (c)Vitamins             (d) Carbohydrates

3.    Carbohydrates are stored in plant body in the form of?

(a)Sugar          (b)Starch              (c)Fats        (d)Glycogen

4.    The food we eat is stored in our body in the form ?

(a)Starch          (b) Glycogen          (c) Fats        (d) Sugar

5.    The molecular formula of glucose is?

(a)$C_6H_{12}O_6$        (b) $C_6H_2O_2$    (c)$C_6H_{12}O_6$    (d)$C_6H_3O_5$

6.    The molecular formula of sugar is?

(a) $C_6H_{12}O_6$      (b) $C_{12}H_{22}O_{11}$        (c) $C_{12}H_{21}O_9$        (d)$C_9H_{22}O_{11}$

7.    Which event is not occurring during Photosynthesis?

(a) Aborsoption of chlorophyll by light energy.

(b) Spiliting  of water molecules .

(c) Conversion of light energy into chemical enrgy.

(d) Reduction of carbon dioxide into carbohydrates.

8.    Plants takes which gas of respiration?

(a)$CO_2$            (b)$O_2$          (c)CO            (d)MgO

9.    _____ contains chlorophyll?

(a)Xylem          (b) Chloroplast      (c) Phloem   (d) Both (a) & (b)

10.   The tiny pores present on the surface of leaves are known?

(a)Poral          (b) Stomata          (c) Stomach          (d)Both (b)&(c)

11.   The opening and closing of the pore is a function of?

(a)Stomata      (b) Guard cell(c)Stomatal pore  (d)None of these

12.   For the synthesis of protein and other compound which metal is

essential?

(a)$NO_2$       (b) N       (c)$CO_2$       (d)$O_2$

13. Which bacteria are useful for nitrogen fixation?

(a)Rizopus       (b)Rizoed       (c)Rizobium       (d)Rizobesilus

14. The alimentary canal is about?

(a)6m       (b)9m       (c)9Km       (d)8Km

15. The small intestine is about?

(a)6m       (b)6.5m       (c)6.3       (d)6.6

16. Saliva is secreted by?

(a)Teeth    (b) Salivary gland    (c) Tongue    (d)Both(a)&(b)

17. Saliva makes the _____

(a)Passage rough       (b) Passage smooth

(c)Passage wet       (d) Both(b)&(c)

18. The movement that push the food forward to the oesophagus?

(a)Peristaltic movement       (b)Peritaltic movement

(c)Periataltic movement       (d)None of these

19. Mucus protect_____ by the action of HCl?

(a)Inner lining of the stomach (b)Outer lining of the stomach

(c)All of the body       (d)Both (b)&(c)

20. Which gland release pepsin?

(a)Bile duct       (b) Gastric gland

(c)Gustatory gland       (d) Pancreatic gland

21. The exit of food from the stomach is regulated by?

(a)Contract muscles       (b) Peristaltic muscles
(c)Sphincter muscles       (d)Stomachal muscles

22. Which is the longest part of the alimentary canal?

    (a)Large intestine            (b) Small intestine

    (c)Stomach            (d) Anus

23. Which is the site of the complete digestion of carbohydrates,fats

    and proteins?

    (a)Stomach     (b) Small intestine   (c) Large intestine      (d) Anus

24. The enzyme pepsin work well in?

    (a)Alkaline medium            (b)Acidic medium

    (c)Neutral medium            (d)Basic medium

25. The pancreatic enzyme work well in_____?

    (a)Acidic medium            (b)Acid medium

    (c)Both(b)&(c)            (d)Alkaline medium

26. Bile juice is _____ in nature?

    (a)Acidic            (b) Both(b)&(c)

    (c)Alkaline            (d) Neutral

27. Enzyme like trypsin and lipase is secreted by?

    (a)Intestinal juice            (b) Pancreas

    (c)Gastric gland            (d)Bile duct

28. The inner lining of the small intestine has numerous fingers like

    Projection called?

    (a)Valli            (b)Villi

    (c)Mount            (d)Velli by anus

29. The exit of the waste material is regulated by ?

(a)Anus                                    (b) Anal sphincter

(c)Anal                                    (d) Anus sphincter

30.    Dental caries or tooth decay causes softening of ?

(a)Dentine                              (b) Enamel

(c)Tooth                                 (d) Both (a)&(b)

31.    Saliva is _____ is nature

(a)Acidic                                (b) Neutral

(c)Basic                                 (d) Both (b)&(c)

32.    The breakdown of glucose into pyruvate takes place in ?

(a)Cell                                   (b)Stomael

(c)Tissue                                (d)Cytoplasm

33.    The breakdown of pyruvate into $co_2$ ,water and energy takes place in ?

(a)Yeast                                 (b)Muscle

(c)Lysosome                           (d) Mitochondria

34.    _____ is the energy currency of most cellular respiration?

(a)ADP            (b)ATP                 (c)APP         (d)Both (a) &(b)

35.    Which provide a surface area for the exchange of gasses?

(a)Nostrils       (b) Alveoli    (c) Larynx    (d)Trachea

36.    The respiratory pigment present in the human body is?

(a)Haemoglobin (b)Tryspin (c)Chloroplast  (d)Cytokinnin

37.    The blood pressure is measured by?

(a)Sphygometer                     (b)Shygomomanometer

(c)Sphygorameter                   (d)Bloodpremeter

38. The si unit of blood pressure is?

    (a)N          (b)Pa       (c)Pascal          (d)Both (b)&(c)

39. Which can avoid the leakage of blood?

    (a)Platelet     (b)Capillaries     (c)Aorte     (d)Both (b)&(c)

40. The transport of water and animals from soil to plant body is done by?

    (a)Lympta       (b) Lymph    (c)Xylem     (d)Phloem

41. The transport of blood from leaves to other part of the plant is done by?

    (a)Lymph       (b)Xylem     (c)Phloem        (d)Lympth

42. Which of the following is not the part of the nephores?

    (a)Bowman's capsule          (b)Glomerulus

    (c) Collecting duct           (d) None of these

43. Which of these following is not the part of digestive system ?

    (a)Trachea      (b) stomach         (c) Food pipe      (d)None of these

44. Which is the power house of cell?

    (a)Lysosome           (b) ATP

    (c)Mitochondria        (d) Golgi apparatus

45. The kidney is the part of?

    (a)Respiration        (b)Excretion      (c)Transportation (d)Nutrition

46. The autotrophic mode of nutrition requires?

    (a)$CO_2$        (b) All of these      (c) Chlorophyll     (d)Sunlight

47. The breakdown of the pryuvate to give ethanol takes place?

    (a)Muscles     (b) Mitochondria       (c)Lysosome      (d) Yeast

48. The anus in human beings in a part of?

(a)Transportation system (b)Alimentay canal

(c)Respiratory system     (d) Excretion system

49.     By which process amoeba takes food from the outside?

(a)Pseudopodia        (b)Endrecytosis     (c)Exocytosis        (d)Both (a)&(b)

Answers:

| QUES. | ANS. | QUES. | ANS. | QUES. | ANS. | QUES. | ANS. | QUES. | ANS. |
|-------|------|-------|------|-------|------|-------|------|-------|------|
| 1 | C | 11 | B | 21 | C | 31 | C | 41 | C |
| 2 | C | 12 | B | 22 | B | 32 | D | 42 | D |
| 3 | B | 13 | C | 23 | B | 33 | D | 43 | A |
| 4 | B | 14 | B | 24 | B | 34 | B | 44 | C |
| 5 | C | 15 | B | 25 | D | 35 | B | 45 | B |
| 6 | B | 16 | B | 26 | C | 36 | A | 46 | B |
| 7 | A | 17 | D | 27 | B | 37 | B | 47 | D |
| 8 | B | 18 | A | 28 | B | 38 | D | 48 | B |
| 9 | B | 19 | A | 29 | B | 39 | A | 49 | B |
| 10 | B | 20 | B | 30 | D | 40 | C | 50 | |

# 6. CONTROL AND CO-ORDINATION

## SOME IMPORTANT POINTS

➢ All information from our environment is detected by the specialised tips of nerve cells called receptors. These are usually our sense organ.

➢ The gap between two neuron is synapse.

➢ Reflex action is response without thinking or any conscious thought. This response is almost involuntary.

➢ The pathway involved in a reflex action is called a reflex arc.

➢ Spinal cord is a cylindrical structure and it about 45cm long.

➢ Human brain has 3 parts :

1. Fore brain: It is thinking part of brain.

2. Mid brain: Control reflexes involve eyes and ears.

3. Hind brain: Controls body equilibrium.

➤ Plant cell change shape changing the amount of water in them, resulting in swelling or shrinking.
➤ The movement towards light is phototrophic movement, towards water is hydrotropism, and toward gravity is geotropism.
➤ Chemotropism: It is the direction movement of plant part in response to chemical stimulus.
➤ Plant hormone auxin stimulates to growth and gibberellins promote stem elongation.
➤ Iodine is necessary for thyroid gland for the production of thyroxin hormone.
➤ Insulin regulates blood glucose level.

# 6. CONTROL AND CO-ORDINATION

1.      In animals control and coordination are provided by?

   (a)Nervous system          (b) Muscular tissues

   (c)Both (a) & (b)           (d) None of these

2.      The receptors which detects change in our environment located in our sense

   Organ, these are?

   (a)The nose                 (b) The ear

   (c) All of these            (d)The tongue

3.      The receptors which detect taste are?

(a)Olfactory receptors          (b) Gustatory receptors

(c)Both (a) & (b)               (d) None of these

4.      The receptors which detect smells are?

(a)Olfactory receptors          (b) Gustatory receptors

(c) Both (a) & (b)              (d) None of these

5.      The electrical impulse travels from to the?

(a)Axon                         (b) Cell body

(c)Synapse                      (d) None of these

6.      The delivery of the impulses from neurons to other cells is allowed by?

(a)Axon                         (b) Cell body

(c)Synapse                      (d) None of these

7.      Where information is acquired first in a neuron?

(a)Axon                         (b) Dendrite

(c)Synapse                      (d) None of these

8.      The electrical impulse sets of the release at some chemical signals at?

(a)Axon                         (b) Dendrite

(c)Synapse                      (d) None of these

9.      Nerves from all over the body meet in a bundle in the?

(a)Brain                        (b) Neuron

(c)Spinal cord                  (d) none of these

10.     Reflex arc are generally formed in the?

(a)Brain                        (b) Neuron

(c)Spinal cord                  (d) none of these

11.    Which is a type of receptor of a body?

(a)Muscle                          (b) Skin

(c)  Foot                          (d) none of these

12.    Which is a type of effectors of a body?

(a)Muscle                          (b) Skin

(c)  Foot                          (d) none of these

13.    The central nervous system consists of?

(a)Brain                           (b) Spinal cord

(c)  Heart                         (d) Both (a) & (b)

14.    Which is an example of voluntary action?

(a)Sneezing                        (b) Clapping

(c)Coughing                        (d) none of these

15.    Which is an example of involuntary action?

(a)Sneezing                        (b) Clapping

(c)Talking                         (d) none of these

16.    The communication between the central nervous system and other parts of s body is facilitated by?

(a) Pripheral nervous system (b) Perial nrvous system

(c) Perinial nervous system    (d) None of these

17.    The cranial nerves and the spinal nerves are consist of?

(a)Pripheral nervous system (b)Perial nrvous system

(c)Capillary                       (d)   None of these

18.    The sensory impulses from various receptors are received by?

(a)Fore brain                      (b) Mid brain

(c)Hind brain             (d) None of these

19.     The main thinking part of the brain is?

(a)Fore brain             (b) Hind brain

(c)Mid brain             (d)None of these

20.     The decision hearing, movement of voluntary muscles, smelling, sight, and so

on, are specified by ?

(a)Fore brain             (b) Mid brain

(c)Hind brain             (d) None of these

21.     A part of the hind brain is?

(a)Cerebrum             (b) Cerebellum

(c)Capillary             (d) Vertebral

22.     The involuntary action including blood pressure, salivation and vomiting are

Controlled by?

(a)Fore brain             (b)Mid brain

(c)Hind brain             (d)None of these

23.     The part of brain which is responsible for voluntary actions?

(a)Cerebrum             (b)Cerebellum

(c)Capillary             (d)Medulla

24.     Which fluid acts as a shock absorbe in brain ?

(a)Andruistic fluid             (b)Paraphenocarpy

(c)Cerebrospinal fluid       (d) Adrenaline

25.     The spinal cord protected by?

(a)Glliberllins             (b)Vertebral column

(c)Parathyroid            (d) Gonads

26.        Which is a sensitive plant?

(a)Mimosa's family            (b)Bryophyllum

(c)Pisum Satvium            (d)None of these

27.        Which part of sensitive plant involve in response to touch?

(a)Stem            (b) Root

(c)Leaves            (d)None of these

28.        In phototropic movements plants respond towards?

(a)Gravity            (b)Light

(c)Air            (d)Heat

29.        In geotropism movements plants respond towards?

(a)Gravity            (b)Light

(c)Air            (d)Heat

30.        In hydrotropism movements plants respond towards?

(a)Gravity            (b) Light

(c)Water            (d) Heat

31.        In chemotropism movements plants respond towards?

(a)Gravity            (b)Light

(c)Air            (d)Chemical

32.        Which hormone synthesized at the shoot tip of plant?

(a) Auxin            (b) Cytokines

(c)Gibberellins            (d)Abscisic acid

33.        Which harmone help in promoting growth in plants?

(a)Auxin          (b) Cytokines

(c)Gibberellins          (d)Abscisic acid

34. Which harmone inhibits growth in plants?

(a ) Auxin          (b) Cytokines

(c) Gibberelllins          (d) Abscisic acid

35. Which hormone synthesized at the stem of plant?

(a)Auxin          (b)Cytokinns

(c)Gibberelllins          (d)Abscisic acid

36. Which hormone causes increasing in heart beats?

(a)Insulin          (b) Adrenaline

(c)Thyroxin          (d)None of these

37. The gland for which iodine is necessary?

(a)Adrenal gland          (b) Pituitary gland

(c)Thyroid gland          (d)None of these

38. Which is a type of emergency gland?

(a)Adrenal gland          (b)Pituitary gland

(c)Thyroid gland          (d)None of these

39. The hormone Which is responsible for growth in animals ?

(a)Insulin          (b) Thyroxin

(c)Adrenaline          (d) Pituitary

40. One of the symptoms of goitre is ?

(a)Increasing fat          (b) Unexpected growth

(c) Swollen neck          (d)None of these

41. The goitre disease is due to deficiency in ?

   (a)Protien                     (b)Iodine

   (c)Carbohydrate                (d)Energy

42. The growth hormones are secreted by ?

   (a) Adernal gland              (b)Pituitary gland

   (c)Thyroid gland               (d)Pancrease gland

43. The hormone which regulating blood sugar levels is?

   (a)Thyroxin                    (b)Adernaline

   (c)Insulin                     (d)Estrogen

44. The insulin hormone is secreted by?

   (a)Adernal gland               (b)Pituitary gland

   (c)Thyroxin gland              (d)Pancrease gland

45. The hormone which is released which by testis gland?

   (a)Oestrogen                   (b)Testostrone

   (c)Thyroxin                    (d) Insulin

46. Which harmone promote fruit ripening oin plants ?

   (a) Auxins                     (b) Gibberellins

   (c) Cytokinns                  (d) Ethylene

47. The gap between two neurons is called a?

   (a)Dendrite                    (b) Synapse

   (c)Impulse                     (d) Axon

48. Which gland is an endocrine glands ?

   (a) Pancreas                   (b) testis

(c)Both (a)&(b)                    (d)None of these

49.    Which is not a plant?

       (a)Auxin                    (b)Oestrogen

       (c)Cytokinns                (d)None of these

50.    The breathing rate increases due to increasing at?

       (a)Diaphragm               (b) Rib muscles

       (c)Both (a) & (b)          (d) None of these

Answers:

| Q. | A. | Q. | A. | Q. | A. | Q. | A. | Q. | A. | Q. | A. | Q. | A. | Q. | A. | Q. | A. | Q. | A. |
|----|----|----|----|----|----|----|----|----|----|----|----|----|----|----|----|----|----|----|----|
| 1 | C | 6 | C | 11 | B | 16 | A | 21 | B | 26 | A | 31 | D | 36 | B | 41 | B | 46 | D |
| 2 | C | 7 | B | 12 | A | 17 | A | 22 | C | 27 | C | 32 | A | 37 | C | 42 | B | 47 | B |
| 3 | B | 8 | A | 13 | D | 18 | A | 23 | B | 28 | B | 33 | B | 38 | A | 43 | C | 48 | C |
| 4 | A | 9 | C | 14 | B | 19 | A | 24 | C | 29 | A | 34 | D | 39 | D | 44 | D | 49 | B |
| 5 | B | 10 | C | 15 | A | 20 | A | 25 | B | 30 | C | 35 | C | 40 | C | 45 | B | 50 | C |

# 7. HOW DO ORGANISM REPRODUCE

## SOME IMPORTANT POINTS

- Reproduction is necessary to continuity on life on Earth.
- The chromosomes in nucleus contain the information for the inheritance of features for parent to offspring in form of DNA.
- The basic event is creating the DNA copy
- DNA copies generated similar not be identical.
- Variations are thus useful for survival of species.
- Fission: It is the mode of asexual reproduction.
- It is two types:
- Binary Fission: The division of parent cell into identical to daughter cell.
- Multiple Fission: The division of parent cell into many individual.
- Leis mania reproduce by binary fission. It causes a disease called Kala-azar.

- Fragmentation: The process of breaking the parent cell into fragments called fragmentations. It is also asexual reproduction.
- Regeneration: Planaria and Hydra like animal regenerate its loss body parts in by injury or autonomy.
- The production of new individual from a out growth of the parent body due to the repeated cell division at a specific site.
- Reproduction takes place by the vegetative parts of plants (leaves, roots, stem) is called vegetative propagation.
- Spore formation: A spore is a single or a several reproductive celled structure that detaches from the parent and under suitable condition germinates new plants.
- Sexual mode of reproduction.
- The fusion of male and female gametes is called fertilisation.
- In flowers male reproductive part is stamen consists filament, anther and female reproductive part pistil consists stigma, style, and ovary.
- Self pollination: Transformation of pollen grain from anther to flower to stigma of some flower or another flower of same plant.
- Cross pollination: Transformation of pollen grain from anther of flower to stigma of another flower of a plant of same species.
- Fertilisation in the human takes place in the fallopian tube.
- Testes produce sperms and male hormone testerone.
- The embryo get nourishment inside the mother body through placenta.
- Sexually transmitted diseases (STD$_s$) are :
- Virus diseases: HIV-AIDS and Warts.
- Bacterial diseases: Syphill and gonorrhoea.

## 7. HOW DO ORGANISMS REPRODUCE

1.      What is need of reproduction?

        (a)For survival                          (b) Continuity of life on earth

        (c)For circulation of blood          (d)None of these

2.      Where are chromosomes present in cell?

        (a)Nucleus       (b)Mitochondria   (c)Cytoplasm        (d)Ribosome

3.      What is the standard form of DNA ?

        (a)Deoxy Nucleic Acid         (b)Deoxyribo Nucleic Acid

        (c)Deoxygenerated Acid      (d)Deoxygenreted Nulcleic Acid

4.      Who is the information source for making proteins ?

(a)Cytoplasm   (b) RNA   (c) DNA   (d)None of these

5.   Which of the following factor leads to variation?

(a)Temperature   (b)Water levels can vary

(c)Meteorite hits   (d)All of these

6.   Which of the following mode of the sexual reproduction ?

(a)Vegitative Propagation   (b)Fission   (c)Spore formation   (d)All of these

7.   Which of the following reproduced by binary fission ?

(a)Leishmania   (b)Yeast   (c)Both (a)&(b)   (d)Spirogyra

8.   Which disease cause by Leishmania ?

(a)Cholera   (b)Kala –azar   (c)Hypatits –B   (d)Elphatatis

9.   Which of the following organism reproduce by multiple fission ?

(a)Sipogyra   (b)Leishmania   (c)Plasmodium   (d)Amoeba

10.   Which of the following of the organism by spore formation ?

(a)Spirogyra   (b)Leishmania   (c)Plasmodium   (d)Amoeba

11.   Hydra is reproduced by?

(a)Budding   (b)Regenration   (c)Both (a)&(b)   (d)None of these

12.   Which of the following reproduced by vegetative propagation?

(a)Rose   (b)Banana   (c)Both (a)&(b)   (d)Papaya

13.   Jasmine is reproduced by?

(a)Layering   (b)Cutting   (c)Grafting   (d)None of these

14.   Rose in reproduced by ?

(a)Layering   (b)Cutting   (c)Grafting   (d)None of these

15.   Mango is reproduced by ?

(a)Layering    (b)Cutting   (c)Grafting       (d)None of these

16.      In which of the following gamets are formed?

        (a)Fission              (b)Regenration

        (c)Spore formation    (d)Sexual reproduction

17.      Rihzopius reproduced by ?

        (a)Budding     (b)Spore formation    (c)Regenration    (d)Fission

18.      What is called the process of fusion and male and female gamets ?

        (a)Self pollination    (b)Cross pollination

        (c)Fertilization        (d)Mensuration

19.      From which part of plant vegetative propagation takes places ?

        (a)Stem       (b)Root     (c)Leaves    (d)All of these

20.      Which of following however is bisexual ?

        (a)Hibscus   (b)Mustard   (c)Both(a)&(b)  (d)Watermelon

21.      Where fertilization takes place in flowering plant body?

        (a)Fallopian tube  (b)Ovary  (c)Pollen tube  (d)None of these

22.      After fertilisation,the zygote divides how many times to form an embryo within the ovule?

        (a) one         (b two       (c) three    (d) several times

23.      Where fertilization takes place in human beings?

        (a)Vasdeferens (b)Fallopian tube (c)Womb (d)Uterus

24.      Which of the following is a party of human male reproductive system ?

        (a)Uterus (b)Cervix (c)Urethra (d)Fallopian tube

25.      Which of the following part of female reproductive system?

        (a)Womb (b)Cervix (c)Urethra (d)Both (a)&(b)

26. Which harmone regulate the production of sperm in males?

(a)Testis (b)Testetrone (c)Exocrine (d)Pancreas

27. Female germ cell is ?

(a)Sperm (b)Womb (c)Egg (d)None of these

28. How many eggs are produced every month in females?

(a)1      (b)2    (c)3   (d)4

29. The time period from fertilization up to birth of the baby is called _____

(a)Germinate period  (b)Gestation period

(c)Both (a)&(b)        (d)Gravitation period

30. Which of the following is sexually transmitted disease?

(a)Warts        (b)Cholera   (c)Pneumonia        (d)Peptic ulcer

31. Standard form of HIV ?

(a)Humanity immunity virus        (b)Human immunity virus

(c)Human immuno virus        (d)None of these

32. Standard form of AIDS ?

(a)Inherited manity syndrome      (b)Aquired imuno deficiency syndrome

(c)Aquired imunity definity syndrome      (d)Nonne of these

33. Which surgical method used in females ?

(a)Vasectomy (b)Sterlization (c)Tubectomy (d)Tuberclosses

34. Which of the following is bacterial STD ?

(a)Syphills (b)Warts (c)Hiv –Aids  (d)Elephantitis

35. Pistil consist?

(a)Stigma, Style                (b) Ovary, fallopian tube

(c)Stigma, Style, Ovary        (d)Petal

36.     Which of the following in female sex harmone ?

(a)Oestrogen  (b)Progestrone  (c)Both (a)&(b)  (d)Testestrone

37.     What happen when fertilization is not occur?

(a)Implantation (b)Mensuration (c)Plantation (d)None of these

38.     The embryo gets nutrition from mothers blood with help of _____

(a)Implantation (b)Mensuration (c)Placenta (d)Gestation

39.     Mensuration  occurs at a regular internal of _____ days .

(a)28      (b)18          (c)20          (d)24

40.     After that the ovaries do not release egg this style is known as _____

(a)Gastation    (b)Menopause     (c)Vascetomy     (d)None of these

41.     which type of reproduction Mucor will reproduced ?

(a)Vegitative Propegation  (b)Budding (c)Fission (d)Spore formation

42.     Pollen tube enters the ovule through a small opening  called_____

(a)Micropyle (b)Ovary (c)Synergids (d)None of these

43.     How many male gamet have one pollen tube?

(a)1      (b)3          (c)2          (d)Either 3 or 2

44.     In human made reproductive system which part provides an optimal

Temperature is 1-3⁰ c lower than normal temperature of the body ?

(a)Vasdeferesns      (b)Testes    (c)Urethra          (d)Scrotum

45.     From which process we can grow seedless plant ?

(a)Pollination                              (b)Regeneration

(c)Vegetative propagation          (d)Spore formation

Answers :

| Q | A | Q | A | Q | A | Q | A | Q | A |
|---|---|---|---|---|---|---|---|---|---|
| 1 | B | 10 | A | 19 | D | 28 | A | 37 | B |
| 2 | A | 11 | C | 20 | C | 29 | B | 38 | C |
| 3 | B | 12 | C | 21 | B | 30 | A | 39 | A |
| 4 | C | 13 | A | 22 | D | 31 | C | 40 | B |
| 5 | D | 14 | B | 23 | B | 32 | B | 41 | D |
| 6 | D | 15 | C | 24 | C | 33 | C | 42 | A |
| 7 | A | 16 | D | 25 | D | 34 | A | 43 | C |
| 8 | B | 17 | B | 26 | B | 35 | C | 44 | D |
| 9 | A | 18 | C | 27 | C | 36 | C | 45 | C |

# 8. HEREDITY AND EVOLUTION

## SOME IMPORTANT POINTS

➢ Inheritance is the result of variation during the process of reproduction.

➢ These variations are essential for survival.

➢ Every individual which are sexually reproduced has two copies of gene. One is dominant and other is recessive.

➢ In offspring of sexual reproduction new combination of traits can be gain.

➢ In different species sex of the offspring is determined by different factors.

➢ In human being sex is determined by the paternal chromosome. Whether it is X (girl) or Y (boys).

➢ Variation can be for only genetic drift or survival advantage or both.

➢ The changes which are due to environment factor on the non reproductive tissue do not affect inheritance.

➢ For a new species genetic drift as well as separation is essential.

- For classifying organism the evolutionary relationship study is done.
- For evolutionary studies fossils are good evidence.
- Complex organs are the result of survival and modification of simple one.
- Different body structures have been adopted and modified as per their functional use and modification in environment.
- Evolution doesn`t signify the progress from simple organism to complex. Evolution both type of organism flourishes at the same time.
- Human being evolved in Africa and then spread all over the world.

## 8. HEREDITY AND EVOLUTION

1.   The number of successful variation are maximised by the process of?

   a. Sexual reproduction      b. Asexual reproduction

   c. Both (a) & (b)           d. None of these

2.   Inheritance from the previous generation provides both a common basic?

   a. Body design  b. Organ design    c. Mental design    d. None of these

3.   If one bacterium divides and then the resultant two bacteria divide again ,the

   four individual bacteria generated would be very similar. There would be only

   very minor differences between them , generated due to only very minor

   differences between them, generated due to small in accuracies in _____ copying?

a. RNA   b. ATP      c. DNA      d. None of these

4.      All these variation in a species have equal changes of surviving in the

environment?

a. Depending upon on the nature of variation   b. Due to RNA

c. Both (a) & (b)                             d. None of these

5.      Bacteria that can with stand heat will survive better in a?

a. Sound wave      b. Heat wave      c. Light wave      d. All of these

6.      Ghanshyam is a male who`s son ear is same  as ghanshyam .It is due to

Inherited traits ? choose the correct statement :

a. Ghanshyam and his grand son looks same

b. Ghanshyam son`s is looks same as friend of grand son

c. Both (a)&(b)

d. None of these

7.      Each traits can be influenced by both parental and?

a. Maternal DNA      b. Uncle`s aunty DNA      c. RNA      d. None of these

8.      A mendelian experiment consisted of breeding tall pea plants bearing violet

flowers with short pea plants bearing white flowers . The prognancy all bore

violet flowers but almost half of them were short this suggested that the genetic

make up of the tall parent can be depicted  as ?

a. TTWW      b.TTWw      c. TtWw      d. TtWw

9.      Mendel used a number of contrasting visible characters of garden pears?

a. Round/Wrinkled seeds    b. Tall / Short plants

c. White /Violet flowers          d. All of these

10.        The information source for making proteins in the cells is?

           a. Cellular DNA          b. RNA          c. Nucleus    d. ATP

11.        A section of DNA the provides information for one proteins is caused?

           a. DNA    b. Nucleus    c. Vacuole    d. Gene

12.        Plant height can thus depends on the amount of a particular plant?

           a. Hormone      b. Cell          c. Tissue      d. All of these

13.        If the gene for that enzyme has an alteration that makes the enzyme?

           a. Less efficient        b.Gain efficient      c. Both (a) & (b)    d. None of these

14.        Due to enzyme less efficient less the plant will?

           a. Long          b. Tall          c. Short        d. All of these

15.        In the experiment of Mendel each pea plant must have _____ sets of genes

           a. Three        b. Two          c. Four          d. Paired

16.        In Mendelium experiment for this mechanism to work , germ cell must have ____

           Gene set ?

           a. Million        b. Two          c. One          d. None of these

17.        The fact that each gene set is present not as a single long thread of DNA  as

           Separate independent pieces each equal a?

           a. Chromosome          b. Genes      c. Bacteria          d. DNA

  18.      Choose the correct statement :

           a. Each cell will have two copies of each chromosome

           b. DNA work same as RNA

           c. Plant will short due to less enzyme

d. Both (a)&(b)

19. _____ sexes participants in sexual reproduction.

    a. One   b. Three      c. Two        d. Four

20. _____ can change sex .

    a. Snakes       b. Bees       c. Seals        d. Snails

21. In some type of animal`s eggs _____ decide.

    What type of sex born male or female?

    a. Atmosphere        b. Temperature      c. Egg layer        d. None of these

22. Which of the living beings, the sex of the individuals is layerly genetically

    Determined ?

    a. Insects       b. Mammals        c. Human        d. None of these

23. _____ inherited from our parents decide whether we will be boys or girls .

    a. Chromosomes       b. Genes      c. Cells        d. Cell wall

24. Women have perfect pair of sex chromosomes caused ?

    a. XX      b. XY        c. YX        d. All of these

25. Man`s sex chromosomes is?

    a. XY      b. XX        c. XZ        d. None of these

26. When X chromosome is shared by mother the born baby is?

    a. Girl   b. Boy       c. Dog        d. Cat

27. There is an un built tendency to variation during reproduction, both because

    Of error in?

    a. DNA   b. RNA        c. Ceus        d. Bacteria

28. Evolution depends upon?

A.Naturaly selected   b. Accidental selected c. None of these d. Both (a)&(b)

29.     The causes of beetle `s decreasing in population is?

a. Change in the DNA of germ cells        b. Because of starvation

c. Heavy rain                              d. None of these

30.     A cross between two plants having two pairs of contrasting is caused?

a. Monohybrid        b. Dihybrid        c. Sexualbrid        d. All of these

31.     In human beings there are pairs of chromosomes?

a. 23      b. 22        c. 24        d. 20

32.     Out of these 22 chromosomes pairs called?

a. Chromosome        b. Cells        c. Autosomes        d. All of these

33.     The last pair of chromosomes that help in deciding gender of that individual is

Called?

a. Genes        b. Nucleus        c. Chromosomes        d. All of these

34.     XX chromosomes present in?

a. Girls        b. Boys        c. Trasents        d. None of these

35.     Branch of science that deals in heredity and variation called?

a. Heredity        b. Variation        c. Genetics        d. Physics

36.     The transmission of features /characters traits from one generation to the

next?

a. Genetics        b. Heredity        c. Variation        d. None of these

37.     The differences among the individuals of a species /population called?

a. Varaition        b. Heredity        c. Development        d. All of these

38.     The scientist who stated his experiment on plant breeding and hybridisation

Was named?

a. Albert einstien  b. George Michel  c. Gregor Johnan Mendel  d. None of these

39.     The organ that have different origin and structure plan but some function

called ?

a. Organ system          b. Analogous system          c. Fossils      d. None of these

Answers :

| Q | A | Q | A | Q | A | Q | A | Q | A | Q | A | Q | A | Q | A |
|---|---|---|---|---|---|---|---|---|---|---|---|---|---|---|---|
| 1 | A | 6 | A | 11 | D | 16 | B | 21 | B | 26 | A | 31 | A | 36 | B |
| 2 | A | 7 | A | 12 | A | 17 | A | 22 | C | 27 | A | 32 | C | 37 | A |
| 3 | C | 8 | D | 13 | A | 18 | D | 23 | B | 28 | D | 33 | C | 38 | C |
| 4 | A | 9 | D | 14 | C | 19 | C | 24 | A | 29 | B | 34 | A | 39 | B |
| 5 | A | 10 | A | 15 | B | 20 | D | 25 | A | 30 | B | 35 | C | | |

**NOTES**

www.ingramcontent.com/pod-product-compliance
Lightning Source LLC
Chambersburg PA
CBHW080623180526
45168CB00007B/3030